The Big Encyclopedia of Triassic Animals

Volume I
By
Stanton F. Fink

Renascentia Phoenix

Acknowledgments
and Dedication

To my father, in whose books I discovered my first monsters.

To Will Caligan, whose help and encouragement is one of the primary reasons for this book's existence.

To Mariano Silvera, who should have had his own artbooks

To Doctor David Morafka, who helped teach me to be more picky with my information.

To Thomas Hegna, whose assistance and support continue being of incalculable importance.

To Nikolas Draper-Ivey, he who wields the sword of Ragnarok, and who has been awarded the Decorations of Omega; he who stands tall to cleave open the Heavens.

To David Jaxon, my eventual partner in crime.

To my friends, who helped push me to make this.

Table of Contents

Acknowledgments	page 2
Introduction	page 5
Glossary	page 6
1. *Anisostoma suessi*	page 8
2. *Arcestes pacificus*	page 10
3. *Austrolimulus fletcheri*	page 12
4. *Elephantosaurus jachimovitschi*	page 14
5. *Eucoelophysis baldwini*	page 16
6. *Foreyia maxkuhni*	page 18
7. *Ophiceras himalayanum*	page 20
8. *Paliguana whitei*	page 22
9. *Progonionemus vogesiacus*	page 24
10. *Shringasaurus_indicus*	page 26
11. *Smilosuchus gregori*	page 28
12. *Thalattosaurus alexandrae*	page 30
13. *Ticinepomis peyeri*	page 32
Bibliography	page 42
About the Artist	page 45

Introduction

This book is an artistic portfolio I've put together of various prehistoric animals from the Triassic Period. The art here are pieces done between 2016 and 2019, and consist of some of the pieces I am most please with, proud of, or otherwise my various favorites for various reasons. Also, since this is a portfolio of lineart, this portfolio also serves as a coloring book for the reader's added personal enjoyment.

Furthermore, I, the artist claim copyright of all art featured in this book, though I officially grant permission for personal or non-commercial use copies to be made. Other inquiries can be made to me via either apokryltaros@gmail.com or stanton.fink@protonmail.com

Glossary

- **Aquatic**- Living in water.
- **Arthropod**- Any member of the animal phylum Arthropoda, including trilobites, arachnids, crustaceans, insects, myriapods and their relatives. All arthropods have armor-like, jointed exoskeletons made of chitin-derived plates, sometimes reinforced with calcium carbonate, and jointed limbs.
- **Cambrian**- A period of time in the Paleozoic Era from 541 to 485 million years ago.
- **Carboniferous**- A period of time in the Paleozoic Era from 359 to 300 million years ago.
- **Cenozoic**- An era of time in the Phanerozoic Eon from 65 million years ago until now.
- **Chordate**- Any member of the animal phylum Chordata, including sea squirts, lancet fish, and vertebrates (such as lampreys, sharks, tuna, frogs, lizards, chickens, and people). All chordates have, at least at some point in their life cycle, a notochord, a long, flexible rod, usually made of cartilage, or, in the case of most vertebrates, cartilage and bone, running down the back from head to tail, directly beneath the neural tube.
- **Cnidarian**- Any member of the animal phylum Cnidaria, such as jellyfish, box jellies, Portuguese Man'o'war, sea anemones, coral and the parasitic myxozoans. Cnidarians are usually radially symmetrical, and have unique, venom-injecting stinging cells called "cnidocytes."
- **Cretaceous**- The last period of time in the Mesozoic Era, from 144 to 66 million years ago.
- **Devonian**- A period of time in the Paleozoic Era from 414 to 360 million years ago.
- **Ediacaran**- The last period of time in the Precambrian Eon from 635 to 542 million years ago.
- **Eocene**- A period of time in the Cenozoic Era from 55 to 33 million years ago.
- **Fauna**- In an ecological context, "fauna" refers to the animal components of an ecosystem.
- **Formation**- In a geological or paleontological context, a formation is a group of rock layers.
- **Gnathostome**- A gnathostome is any vertebrate chordate with a moveable jaw (or had an ancestor with one).
- **Holocene**- A period of time in the Cenozoic Era from 12,000 years ago until now.
- ***Incertae sedis***- A Latin phrase literally meaning "uncertain seat." *"Incertae sedis"* is a term in classification used to refer to a species or group whose relationships with related organisms are unclear or poorly defined.
- **Jurassic**- The second period of time in the Mesozoic Era, from 199 to 145 million years ago.
- **Mesozoic**- An era of time in the Phanerozoic Eon from 249 to 66 million years ago.
- **Miocene**- A period of time in the Cenozoic Era from 23 to 5 million years ago.

- **Mollusk**- Any member of the animal phylum Mollusca, including snails, clams, squid, octopuses, tusk shells and chitons. Most mollusks have a calcium carbonate shell, and a toothed, file-like tongue called a radula. All mollusks have a cape-like organ, the mantle, which usually secretes the shell, and houses breathing organs, and a nervous system.
- **Nekton**- Any aquatic animal that lives either entirely or almost entirely in the water column, and relies on its own swimming or propulsion abilities to keep and move itself in and around the water column. Anchovies, porpoises and ichthyosaurs are examples of nekton.
- **Neogene**- The second third of the Cenozoic Era, comprising of the Miocene and the Pliocene periods.
- **Oligocene**- A period of time in the Cenozoic Era from 33 to 23 million years ago.
- **Ordovician**- A period of time in the Paleozoic Era from 484 to 440 million years ago.
- **Paleocene**- A period of time in the Cenozoic Era from 65 to 55 million years ago.
- **Paleogene**- The first third of the Cenozoic Era, comprising of the Paleocene, Eocene, and Oligocene.
- **Paleozoic-** An era of time in the Phanerozoic Eon from 249 to 66 million years ago.
- **Permian**- The last period of time in the Paleozoic Era, the time of "The Great Dying," or most severe of all known extinction events, from 299 to 250 million years ago.
- **Pharynx**- A structure in the throat of many animals located directly behind the mouth or oral chamber. In vertebrates, it often houses breathing structures, like gills.
- **Plankton**- An organism that uses water currents and waterflow to as its primary means of transportation in the water column because it is either too small to move long distances by its own power, or lacks the ability to propel itself entirely. Sargassum seaweed and jellyfish are two varieties of plankton.
- **Pleistocene**- A period of time in the Cenozoic Era from 3 million years ago until 12 thousand years ago.
- **Pliocene**- A period of time in the Cenozoic Era from 5 to 3 million years ago.
- **Quaternary**- The last third of the Cenozoic Era, comprising of the Pleistocene and the Holocene periods.
- **Terrestrial**- Living on land.
- **Triassic**- The first period of time in the Mesozoic Era, from 249 to 200 million years ago.

Name	*Anisostoma suessi*
Phylum	Mollusca
Class	Gastropoda
clade	Eogastropoda
clade	Euomphalina
clade	Euomphaloidea
Family	Euomphalidae
Size	Estimated to be around 3 to 4 meters long
Time Period	Norian epoch of the Late Triassic Period, between 227 and 208 million years ago
Location	Central Transdanubian Range, Austria
Comments	*Anisostoma suessi* (not to be confused with slime mold beetles of the genus *Anisotoma*) is an extinct marine gastropod that lived in what is now Austria during the Norian Epoch of the Late Triassic. The last whorl of *A. suessi*'s shell is modified into a large, kidney shaped brim around the mouth of the shell, so that the entire head of the living animal was, presumedly entirely covered. The purpose of this immense brim remains vague.

Name	*Arcestes pacificus*
Phylum	Mollusca
Class	Cephalopoda
Subclass	Ammonoidea
Order	Ceratitida
Family	Arcestidae
Size	Adult shell up to 3 centimeters in diameter.
Time Period	Carnian to Middle Norian Epochs, Middle to Late Triassic Period, 232 to 221 million years ago.
Location	*Tropites subbullatus* Zone of Shasta County, California.
Comments	*Arcestes pacificus* is a species of small, spherical ammonoid cephalopod from the Middle to Late Triassic of Northern California. Fossils of *Arcestes* are found in Middle to Late Triassic marine strata throughout the world. *A. pacificus* lived near the seafloor, and preyed on small invertebrates and foraminifera. Its shell had constrictions that made it look like it was split into quarters.

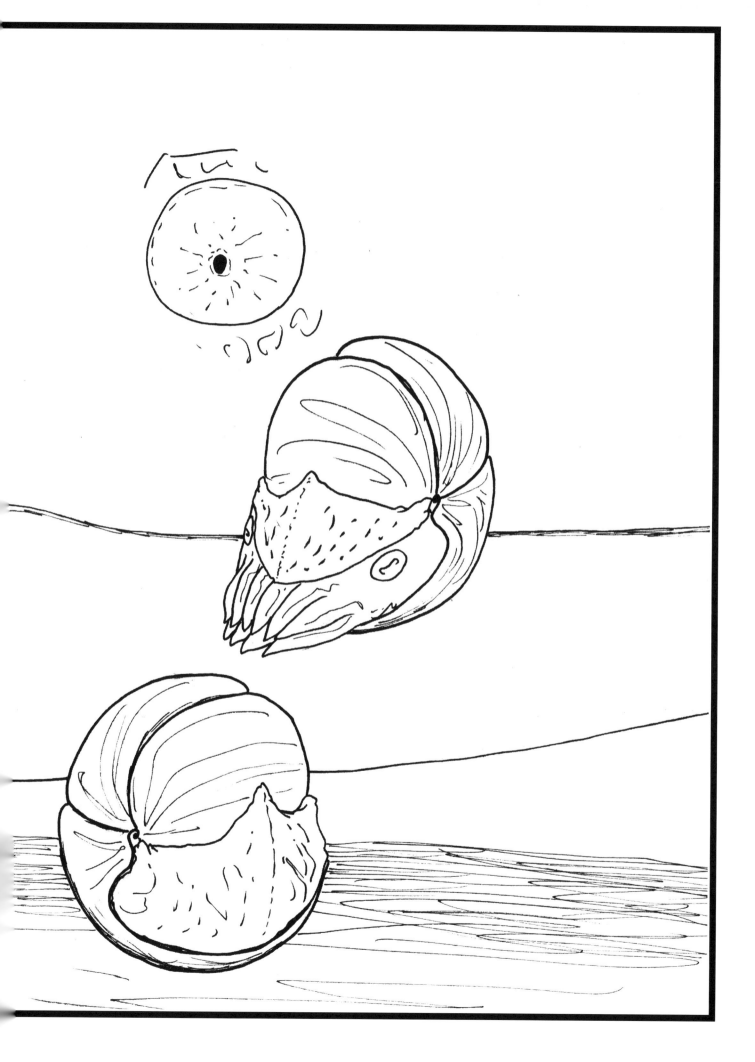

Name	*Austrolimulus fletcheri*
Phylum	Arthropoda
Class	Merostomata
Order	Xiphosura
Family	Austrolimulidae
Size	Length from tip of caudal spine to anteriormost edge 14.6 centimeters; width from genal spine tip to genal spine tip 17.8 centimeters.
Time Period	Ladian Stage of the Middle Triassic, 238 to 240 million years ago.
Location	Beacon Hill shales, in the middle of the Middle Triassic-aged Hawkesbury Series in Brookvale, New South Wales.
Comments	*Austrolimulus fletcheri* is an extinct horseshoe crab from the Middle Triassic of New South Wales. It is best known for having long, sword-like genal spines that, coupled with the long caudal spine, gave the animal the resemblance of a clock pendulum or a miner's pickaxe. In addition to its unusual appearance, arthropod specialists note that *Austrolimulus* represents a transitional form between horseshoe crabs of the family Belinuridae and Limulidae. In horseshoe crabs of the family Belinuridae, the segments of the opisthosoma (the hind-body that corresponds to the abdomen of arachnids, or the body and pygidium of trilobites) are distinct. In horseshoe crabs of the family Limulidae (including the still-living genera of *Limulus* and), the segments of the opisthosoma are fused together entirely into a single, mostly smooth unit. In *Austrolimulus* and its relatives, the segments of the opisthosoma are fused together, but still remain distinct.

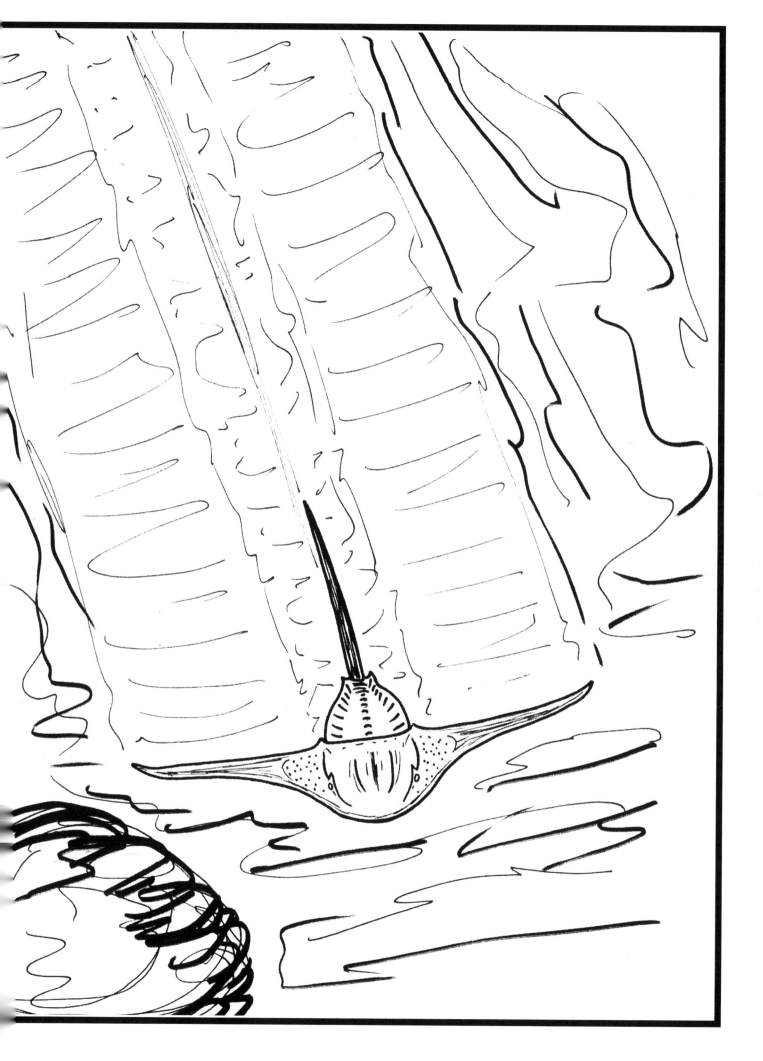

Name	*Elephantosaurus jachimovitschi*
Phylum	Chordata
clade	Sphenacodontoidea
Order	Therapsida
clade	Dicynodontia
clade	Kannemeyeriiformes
Family	?Stahleckeriidae
Size	Skull fragment suggests a large skull at least 30 centimeters wide, living animal may have been up to 3 meters long.
Time Period	Ladian Stage of the Middle Triassic, 238 to 240 million years ago.
Location	Starokoltaevo I, Ural Mountains, Russia
Comments	*Elephantosaurus jachimovitschi* is probably a very large dicynodont therapsid vertebrate from the Middle Triassic of what is now Russia, in the Ural Mountains. The holotype and only known specimen is a fragment from the left side of a very large skull that containing the left half of the nosebridge, portion of the left eye socket, and nasal bones. The remains suggest an animal related to the Brazilian *Stahleckeria*, though, the anatomy of *Elephantosaurus*' eye socket makes the species' placement in Stahleckeriidae questionable.

Name	*Eucoelophysis baldwini*
Phylum	Chordata
Class	Reptilia
clade	Archosauria
clade	Dinosauriformes
clade	Dracohors
Family	Silesauridae
Size	Pubis bone up to 20 centimeters long
Time Period	Revueltian faunachron of the Early Norian Epoch, Late Triassic Period, 227 to 225 million years ago.
Location	Cross quarry on Orphan Mesa, Petrified Forest Formation of the Chinle Group, north-central New Mexico.
Comments	*Eucoelophysis baldwini* is a species of silesaurid dinosauriform from Late Triassic New Mexico. *Eucoelophysis* (and other members of the family Silesauridae) is not a dinosaur *per se*, but is related to the common ancestor of saurischian and ornithischian dinosaurs. The first specimens were thought to be bones of the a dinosaur closely related *Coelophysis*, but, two independent studies published in 2005 and 2006 demonstrated that the bones of *Eucoelophysis* were neither of (a) *Coelophysis* (relative), nor of a dinosaur, proper. Here, the animal is restored as a quadrupedal creature similar in form to *Silesaurus*.

Name	*Foreyia maxkuhni*
Phylum	Chordata
Class	Sarcopterygii
Order	Coelacanthiformes
Family	Latimeriidae
Size	About 190 millimeters long
Time Period	Lower Ladinian Epoch of the Middle Triassic, 491 million years ago
Location	Canton Ticino, Monte San Giorgio, Switzerland
Comments	*Foreyia maxkuhi*, named for Peter Forey and Max Kuhn, respectively, is a strange, aberrant coelacanth from marine strata in what is now the mountains of southern Switzerland along the Italian border. *F. maxkuhni* has a proportionally huge, rounded head with a low, horn-like point, a hooked maxilla and an underbite, possibly adaptations for being a slow-moving grazer of encrusting animals. The coelacanth bodyplan normalized during the Late Carboniferous, and would see few deviations afterwards, besides *Foreyia* and the fork-tailed *Rebellatrix*. Despite its weird appearance, *Foreyia* is a relative of the modern coelacanth, *Latimeria*, and its closest relative is the plainer-looking *Ticinepomis*, whose fossils are found in the same strata and region.

Name	*Ophiceras himalayanum*
Phylum	Mollusca
Class	Cephalopoda
Subclass	Ammonoidea
Order	Ceratitida
Family	Ophiceratidae
Size	Average adult shell diameter up to 4 centimeters wide.
Time Period	Scythian or Griesbachian Epoch of the Early Triassic, 253 to 251 million years ago.
Location	The type locality is Shalshal Cliff, Bed 2, Rimkin Paiar, Northern India.
Comments	*Ophiceras himalayanum* is a species of ceratitid ammonoid from Early Triassic northern India, and is a member of a mostly Early Triassic genus (with the sole exception of *O. bilotense*, which is restricted to the late Permian portions of the Salt Mountains in Pakistan). All of the whorls of *O. himalayanum* are visible, that is, the embryonic and older whorls are not hidden or obscured as the large, younger whorls are produced during the animal's lifetime. In the juvenile, the shell is ornamented with slight folds. As the animal grew older, these folds become more and more pronounced, until they become coarse ribs in the younger whorls of the adult's shell. Although fossils of *O. himalayanum* are found in the Himalayas Mountains, this species is restricted to the Indian Himalayas, and should not be confused with the Tibetan species, *O. tibeticum*, which is found in the Tibetan Himalayas. The genus *Ophiceras* should also not be confused with the similarly named Silurian nautiloid, *Ophioceras*.

Name	*Paliguana whitei*
Phylum	Chordata
Class	Reptilia
clade	Lepidosauromorpha
Order	"Eolacertia"
Family	Paliguanidae
Size	Only known partial skull about 3 centimeters long.
Time Period	Induan Epoch, Early Triassic Period, 252 million years ago.
Location	Katberg Formation in Donnybrook, Queenstown District, South Africa.
Comments	*Paliguana whitei* is a primitive lepidosauromorph reptile known from a damaged, partial skull from what is now Donnybrook, South Africa, dating from the start of the Triassic Period. When *Paliguana* was first described, it was thought to be a true lizard, making it the earliest squamate reptile. Now, however, *Paliguana* is now thought to be a lepidosauromorph probably related to squamates and rhynchocephalians (i.e., the tuatara), and is lumped into the wastebasket taxon "Eolacertia," a group of primitive diapsid reptiles that also includes the gliding Kuehneosauridae. *Paliguana*'s skull has very large eye sockets and may have had very small teeth.

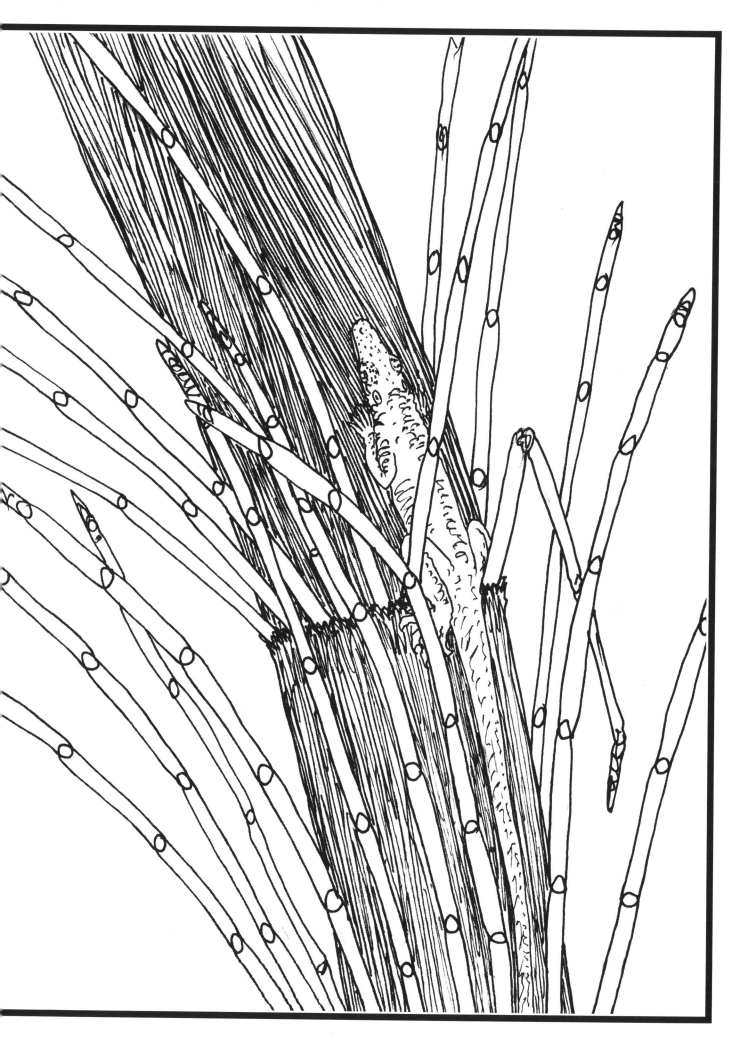

Name	*Progonionemus vogesiacus*
Phylum	Cnidaria
Class	Hydrozoa
Order	Limnomedusae
Family	Olindiidae
Size	Juveniles having bells up to 8 millimeters in diameter, tentacles up to 9 millimeters in length. Adults up to 40 millimeters in diameter, and tentacles up to 40 millimeters in length.
Time Period	Anisian to Ladinian Epochs of Middle Triassic Period, 247 to 242 million years ago.
Location	Grès à Voltzia formation, near Vilsberg, Grand Est, France
Comments	*Progonionemus vogesiacus* is an extinct cnidarian from the Middle Triassic Grès à Voltzia Lagerstätte in northeastern France. The ten or so fossils are of juvenile to adult medusae that bear a striking resemblance to the medusa forms of the limnomedusan hydrozoan genus *Gonionemus*, though *Progonionemus* lacks the modern genus' characteristic elbow-like tentacle suckers. Because *Progonionemus* lacks the suckers of its namesake, *Gonionemus*, the former undoubtedly swam in the water column more frequently. The environment of Grès à Voltzia during the Middle Triassic was a river delta that created numerous series of brackish water ponds in a semiarid scrubland. These ponds would often become deoxygenated, killing all of their inhabitants, right before drying out. The deceased inhabitants, including the ten known individuals of *P. vogesiacus*, would then be preserved with a layer of sediment brought into the dried out ponds both through wind and through floods refilling the ponds.

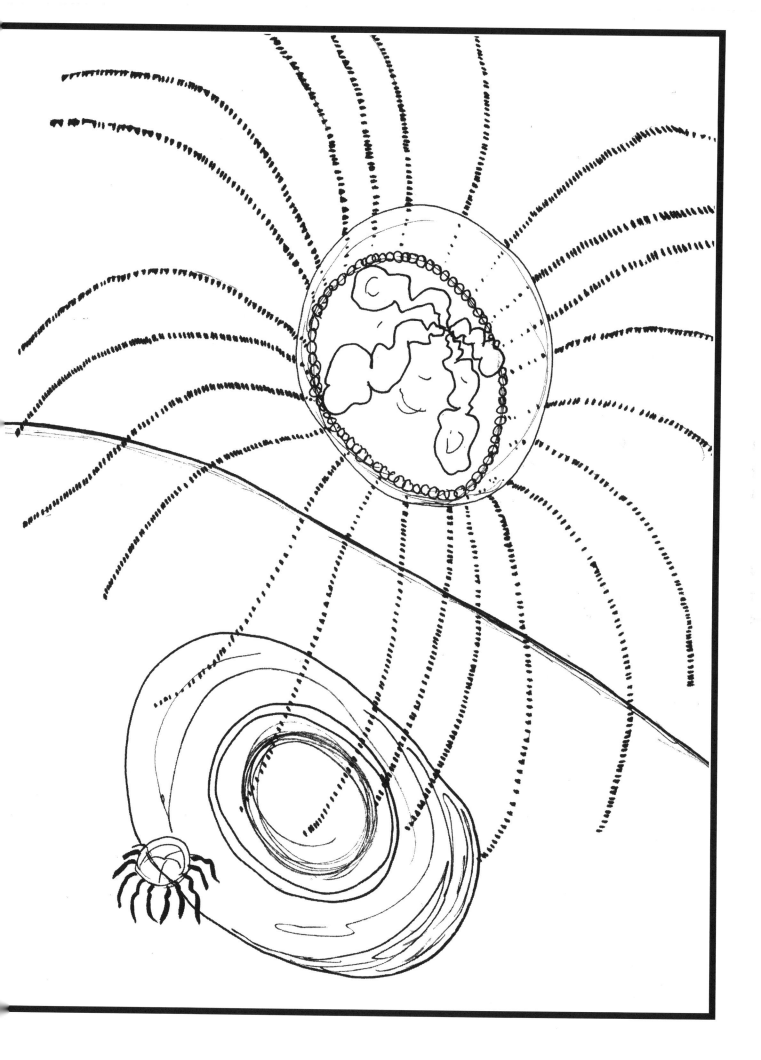

Name	*Shringasaurus indicus*
Phylum	Chordata
Class	Reptilia
clade	Archosauromorpha
clade	Allokotosauria
Family	Azendohsauridae
Size	Estimated to be around 3 to 4 meters long
Time Period	Anisian epoch of the Middle Triassic Period, between 247 and 242 million years ago
Location	Denwa Formation, Madhya Pradesh State, India
Comments	*Shringasaurus indicus* is a horned archosauromorph reptile from the middle Triassic of India. The blocky skull has a pair of large, curved horns that are thought to have been used in displays. The living animal would have been a chunky quadraped up to 3 to 4 meters long. The teeth, which are similar to the teeth of the poorly known Moroccan archosauromorph, *Azendohsaurus*, suggest *S. indicus* was an herbivore. *S. indicus*, or, at least the area where the jumbled bones of the 7 or so known specimens were found, lived in a floodplain, possibly near a lake, with bushy vegetation and a diverse assemblage of animals.

Name	*Smilosuchus gregoriii*
Phylum	Chordata
Class	Reptilia
clade	Archosauriformes
Order	Phytosauria
Family	Phytosauridae
Size	Largest known skull 155 centimeters in length, adult body length estimated to be between 7 and 12 meters
Time Period	Carnian to Norian Epochs, Middle to Late Triassic Period, 235 to 208 million years ago.
Location	Norian-aged specimens from Round Rock Horizon, Nazlini Horizon and "Ward's Bonebeds" of the Chinle Formation, Arizona, and Carnian to Norian-aged specimens from Rotten Hill, Texas.
Comments	*Smilosuchus gregoriii* is a giant phytosaur archosauriform reptile from Late Triassic Arizona and Texas. It had a pair of long, massive jaws modified for both snaring prey and rending flesh. The tusk-like teeth near the tip of the jaws would snag bitten prey, which, similar to prey of modern crocodiles, would then be dragged into the water, allowing the beast to bring the bitten prey closer to the blade-like teeth near the back of the jaws, so flesh and limb could be torn off and swallowed. Phytosaurs strongly resemble modern crocodilians in bodyform and aquatic habits, though, the two can be easily distinguished by how the nostrils of phytosaurs are situated near the eyes, and how the nostrils of crocodilians are at the tip of the snout. Furthermore, Triassic crocodilians were small, lightly built terrestrial animals. The massive rostral crest of *Smilosuchus* caused researchers to believe it to be closely related to the phytosaur genus *Leptosuchus*, which has a similar, but smaller rostral crest. More in-depth studiy of skull anatomy now shows that *Smilosuchus* is more closely related to *Pravusuchus* and the pseudopalatine phytosaurs *Mystriosuchus* and *Machaeroprosopus*, who may be its descendants.

Name	*Thalattosaurus alexandrae*
Phylum	Chordata
Class	Reptilia
clade	Neodiapsida
Order	Thalattosauria
Family	Thalattosauridae
Size	Adult up to 2 meters in length
Time Period	Carnian to Middle Norian Epochs, Middle to Late Triassic Period, 235 to 221 million years ago.
Location	Hosselkus Limestone of Plumas and Shasta Counties, California.
Comments	*Thalattosaurus alexandrae* is an extinct marine reptile from the latter half of the Triassic in northern California. The specific name honors the explorer, naturalist and philanthropist, Annie Montague Alexander. It was a slender animal, superficially similar in build to a marine iguana, but with a more flattened, ribbon-like tail. Adults may have returned to land to sunbathe and lay eggs, similar to marine iguanas. *Thalattosaurus* had a downturned snout, similar to, but larger and more gently curving than that of its close relative, *Nectosaurus* (behind *T. alexandrae*). The downturned snout and jaws are thought to be adaptations for crushing shellfish, such as those of the many ammonites and clams that lived in the Hosselkus Limestone formation.

Name	*Ticinepomis peyeri*
Phylum	Chordata
Class	Sarcopterygii
Order	Coelacanthiformes
Family	Latimeriidae
Size	Up to 18 centimeters long
Time Period	Lower Ladinian Epoch of the Middle Triassic, 491 million years ago
Location	Canton Ticino, Monte San Giorgio, Switzerland
Comments	*Ticinepomis peyeri* is an extinct coelacanth from marine strata in what is now the mountains of southern Switzerland along the Italian border. *T. peyeri* is a relative of both the modern coelacanth, *Latimeria sp.*, and the bizarre *Foreyia maxkuhi*, which also lived with *Ticinepomis* in Canton Ticino. The living animal would have a strong resemblance to the modern coelacanth (as do most Mesozoic coelacanths), differing in having a more elongated snout, and more delicate fins. Detailed examinations of the fossils of *Foreyia* would reveal that it and *Ticinepomis* were closely related, sharing numerous anatomical features despite their dramatically different forms.

Bibliography

- Arkell, W.J. et al., 1957. Mesozoic Ammonoidea. Treatise on Invertebrate Paleontology Part L. Geological Society and University of Kansas Press.
- Broom, R. "On the skull of a true lizard (Paliguana whitei) from the Triassic beds of South Africa." *Records of the Albany Museum* 1.1 (1903): 1.
- Carroll, Robert L. "Permo-Triassic" lizards" from the Karroo." (1975).
- Cavin, Lionel, et al. "Heterochronic evolution explains novel body shape in a Triassic coelacanth from Switzerland." *Scientific reports* 7.1 (2017): 1-7.
- Ferrante, Christophe, et al. "Coelacanths from the Middle Triassic of Switzerland show unusual morphology."
- Grauvogel, Louis, and Jean-Claude Gall. "Progonionemus vogesiacus nov. gen. nov. sp., une méduse du Grès à Voltzia des Vosges septentrionales." *Sciences Géologiques, bulletins et mémoires* 15.2 (1962): 17-27.
- Hyatt, Alpheus, and James Perrin Smith. *The Triassic cephalopod genera of America*. Vol. 40. US Government Printing Office, 1905.
- Kammerer, Christian F., Jörg Fröbisch, and Kenneth D. Angielczyk. "On the validity and phylogenetic position of Eubrachiosaurus browni, a kannemeyeriiform dicynodont (Anomodontia) from Triassic North America." *Plos One* 8.5 (2013): e64203.
- Long, R. A., and Murry, P. A. (1995). Late Triassic (Carnian and Norian) tetrapods from the southwestern United States. *New Mexico Museum of Natural History and Science Bulletin* **4**:1-254.
- LUCAS, SPENCER G., et al. "CARNIAN (LATE TRIASSIC) AMMONOIDS FROM ELANTIMONIO, SONORA, MEXICO." *Fossil Record 4: Bulletin 67* 67 (2015): 189.
- Merriam, John Campbell. *The Thalattosauria: a group of marine reptiles from the Triassic of California*. The Academy, 1905.
- Moore, R. C., et al. 1960. Mollusca 1. Part I of R. C. Moore, ed. Treatise on invertebrate paleontology. Geological Society of America, New York, and University of Kansas Press, Lawrence
- Riek, E. F. "A new xiphosuran from the Triassic sediments at Brookvale, New South Wales." *Records of the Australian Museum* 23 (1955): 281-282.
- Rieppel, O. "A new coelacanth from the Middle Triassic of Monte San Giorgio, Switzerland." (1980).
- Rieppel, Olivier, Johannes Müller, and Jun Liu. "Rostral structure in Thalattosauria (Reptilia, Diapsida)." *Canadian Journal of Earth Sciences* 42.12 (2005): 2081-2086.
- Sengupta, Saradee, Martín D. Ezcurra, and Saswati Bandyopadhyay. "A new horned and long-necked herbivorous stem-archosaur from the Middle Triassic of India." *Scientific reports* 7.1 (2017): 8366.
- Smith, J.P. 1932. Lower Triassic Ammonoids of North America. US Geological Survey Professional Paper 167.
- V'yushkov, V.P., 1969. New dicynodonts from the Triassic of the Cis-Urals. Paleontol. Zh.

1969, 99 – 106

About the Artist

Stanton F. Fink is a student of Biology and Chinese Medicine, and makes a hobby of drawing monsters and researching flowers, arcane-looking creatures, prehistoric animals, fish, reptiles, birds and the occasional, really grotesque fungal fruiting body.

Stanton grew up and went to school in California and is currently living, drawing, and gardening in Oregon.

Made in the USA
Columbia, SC
28 December 2024